A to Z Sharks for Kids

65 Sharks and 65 Unique Illustrations with Interesting and Fun Facts

Professor Smart

Contents

A Thank You Gift

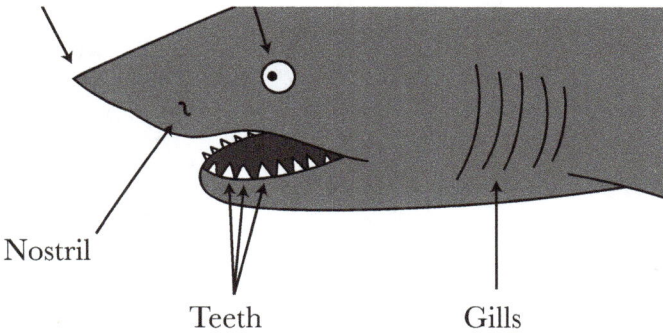

As a token of our appreciation, we would like to share our printable shark anatomy infographic with you! To download this for free, visit bit.ly/shark-infographic or scan the QR code below.

All About Sharks

Sharks have been around longer than dinosaurs.
Scientists have found shark fossils that are 455 million years old, whereas dinosaur fossils are around 230 million years old. It is believed that sharks descended from a leaf-shaped fish that had no eyes or fins.

Sharks can live for a very long time.
Most sharks have an average lifespan of about 20 - 30 years. The oldest known shark in the world is the Greenland shark, which can live up to 400 years old.

Sharks are one of the most agile animals in the ocean.
A shark's skeleton is made up of cartilage, which is stronger, more flexible, and lighter than bone. This allows them to stay afloat, conserve energy, and make sharp turns quickly.

Not all sharks are fast swimmers.
The Shortfin Mako shark can swim up to 60 miles per hour (96 km per hour), whereas the Greenland shark cruises around at 0.76 miles per hour (10.8 km per hour).

Sharks have tiny teeth on their skin.
Sharks have millions of tiny teeth called dermal denticles which are found all over their body. These teeth reduce surface drag and enable the shark to swim very fast. Humans have also replicated this technology in swimming costumes.

Some sharks can lose up to 30,000 teeth in their lifetime.
Shark teeth are made of enamel and are very strong. Sharks are also born with teeth, and they replace their teeth every two weeks. Sharks can have flat crushing teeth, pointed teeth, or sharp serrated teeth.

Sharks use their fins for balance and stability.
Their dorsal fins provide balance, their pectoral fins are used to steer, and their tails are used to propel themselves forward.

Most sharks have five pairs of gill slits.
Sharks use their gills for breathing and must remain in constant forward motion. As they swim, water is passed through their mouth and out of their gills. In this process, oxygen is absorbed into their body through tiny blood vessels.

Two-thirds of the shark's brain is used for their sense of smell.
Sharks use their nose for smelling. They can smell one part of blood for every one million parts of water from hundreds of yards or meters away. They can do this as their nasal sacs are filled with sensory cells, which sends signals to their brain about the smell.

Sharks can see in almost every direction.
Most sharks have their eyes positioned on the sides of their head, and so they can see in almost every direction. However, they have two blind spots - right in front of their snout and just behind the head.

Sharks have excellent hearing.
Sharks have an acute sense of hearing and can also hear very low-frequency signals. Sound travels farther and faster underwater, so sharks can hear their prey up to 800 feet (243 m) away.

Sharks can sense electrical currents in the water.
Sharks have 'ampullae of Lorenzini,' which are special organs near their nose, eyes, and mouth. Shark use these organs to sense electromagnetic fields and temperature changes in the ocean. This helps them to navigate and also find prey.

Sharks can lay eggs or give birth to live pups.
There are three main ways that sharks can reproduce. Oviparous sharks lay eggs that develop and hatch outside of the mother's body. Ovoviviparous sharks carry and hatch eggs inside their body and then give birth to their young. Viviparous sharks develop baby sharks inside their body and give birth to live young or pups.

Sharks live in every ocean on Earth.
Most sharks can be found in oceans, but some also live in rivers. Some sharks, like the bull shark, can live in both saltwater and freshwater.

Depending on their characteristics, sharks belong to one of eight orders of classification.
- The carcharhiniformes are the largest order of sharks. They have the classical shark shape with five pairs of gill slits, two spineless dorsal fins, an anal fin, and a wide mouth with shark teeth.
- The heterodontiformes order only contains nine known species of horn sharks, each with five pairs of gill slits and a dorsal fin with a strong spine. They can have either sharp or flat rounded teeth.
- The hexanchiformes order contains the most primitive sharks that have six or seven pairs of gill slits. They only have one dorsal fin and an anal fin. Their teeth are thorny.
- The lamniformes sharks have the standard five-gill slits and a large mouth with several rows of shark teeth. They have two dorsal fins and an anal fin. They also have a special ability to increase their body temperature to higher than the temperature of the water they are swimming in.
- Sharks in the orectolobiformes order have five pairs of gill slits, two spineless dorsal fins, an anal fin, and spiracles near their eyes. Most of them have patterns on their skin, and some have barbels.
- The pristiophoriformes order contains sawsharks that have long saw-like snouts, five or six-gill slits, two dorsal fins, but no anal fin. Their pectoral fins are wide.
- Sharks in the squaliformes order are found worldwide. About 126 species of sharks belong to this order. They have long snouts, short mouths, five-gill slits, two dorsal fins, and no anal fin.

- The eighth shark order is the squatiniformes. These sharks have flattened bodies, short snouts with dermal flaps, barbels and eyes and spiracles on top of their heads. They don't have an anal fin.

Sharks do not want to eat people.
Sharks often 'test bite' their potential food to see if it's right for them. When sharks bite humans, they usually spit them back out, as they are not interested in eating human flesh. Generally, sharks are not aggressive and will only bite a person if they feel threatened.

The biggest threat to sharks is people.
The International Union for Conservation of Nature (more commonly known as the IUCN) is the authority that reports on the status of the natural world and what needs to be done to safeguard it. The IUCN Red List of Threatened Species reports on the extinction risk status of plant and animal species, including sharks. The listing starts and continues in the following order: Not evaluated > Data deficient > Least concern > Near-threatened > Vulnerable > Endangered > Critically endangered > Extinct in the wild > Extinct.

If we don't protect our sharks, they will become extinct as they can't reproduce fast enough to replace the sharks that die. The biggest threats are overfishing, bycatch, shark finning, habitat destruction, and pollution.

In this book, you will find 65 sharks from A to Z with unique illustrations and interesting, fun facts.

We hope you enjoy this book!

Professor Smart.

AUSTRALIAN BLACKTIP SHARK

Status:	least concern
Maximum size:	5.9 ft (1.8 m)
Lifespan:	20 years
Reproduction:	viviparous
Color on top:	bronze or dark grey
Color underneath:	white to yellow

Features: black tips on every fin except anal fin, pale stripe along the flank, long pointed snout
Habitat: the continental shelf, close to shore
Distribution: northern and eastern Australia

Did you know?
- It is also known as a blacktip whaler.
- It is not the same shark as the common blacktip shark.
- It is caught by commercial fishermen and sold as 'flake' in fish and chip shops.

AUSTRALIAN SWELL SHARK

Status:	least concern
Maximum size:	4.9 ft (1.5 m)
Lifespan:	20 - 35 years
Reproduction:	oviparous
Color on top:	medium greyish brown, irregular light and dark spots
Color underneath:	cream with a dark stripe down the middle

Features: blunt, broad snout with long cat eyes, wide dark saddles behind the eyes, first dorsal fin set way back on the body, pale edges on all fins
Habitat: shallow water above rocky reefs and seaweeds
Distribution: southern Australia

Did you know?
- The Australian swell shark is the most abundant catshark species in southern Australia.
- It can 'swell up' and make itself look bigger by sucking in air and water! It does this to protect itself from larger sharks like the broadnose sevengill shark.
- It is also commonly called the draughtboard shark, flopguts, Isabell's swellshark, and sleepy joe.
- It's a pain in the neck for lobster fishers as it swims into lobster nets looking for food. It is often thrown back into the water by fishermen and survives well out of water.

BAHAMAS SAWSHARK

Status: data-deficient
Maximum size: 2.7 ft (81 cm)
Lifespan: 15 years
Reproduction: ovoviviparous, 17 pups per litter
Color on top: light grey, dark brown stripes along the saw
Color underneath: white

Features: flattened head with a long flat saw studded with teeth, a pair of barbels halfway under the saw
Habitat: deep water off continental shelves
Distribution: western central Atlantic Ocean between Cuba, Florida and the Bahamas

Did you know?

- The Bahamas sawshark is a harmless shark despite its 23 large sawteeth. Thirteen teeth are before the barbels and ten after the barbels.
- Mexico has assessed this shark as endangered or threatened even though there is very little information available about this shark.
- No fisheries are known to be targeting this shark.
- It is suspected that the Bahamas sawshark eats small fish, crustaceans and squids.
- General sawsharks can live in isolation or as part of a large group, although we are still unsure about this exact species.

BASKING SHARK

Status: vulnerable

Maximum size: 40 ft (12 m)

Lifespan: 50 years

Reproduction: thought to be ovoviviparous but very little information is available

Color on top: grey-brown to dark grey or black

Color underneath: slightly paler than the color on top or white

Features: large gill slips circles the head and lunate caudal fin, conical snout and huge mouth

Habitat: coastal waters

Distribution: worldwide, in arctic to temperate waters

Did you know?
- The Basking shark is the second-largest fish, after the whale shark.
- It looks like it's 'basking' in the sun as it swims around slowly and close to the surface with its huge mouth open.
- Despite its size, it is a passive shark and not dangerous for humans.
- People in the past spread tales of sea serpents and monsters when they saw a basking shark.

BIGNOSE SHARK

Status:	data-deficient
Maximum size:	9.8 ft (3 m)
Lifespan:	unknown
Reproduction:	viviparous, 3 - 11 pups per litter
Color on top:	grey
Color underneath:	white

Features: broad, blunt snout, large first dorsal fin, long straight pectoral fins

Habitat: off continental shelves up to 1,410 ft (430 m) deep

Distribution: tropical warm waters, of the Atlantic Ocean, Mediterranean Sea, western Indian Ocean, and the Pacific Ocean

Did you know?

- The Bignose shark is caught by trawlers and used to make fish meal for animals. You're not allowed to catch this shark anymore in US waters.
- It looks like a type of shark called a 'night shark except that it doesn't have a long second dorsal fin free rear tip or green eyes like a night shark.

BLACK DOGFISH

Status:	least concern
Maximum size:	35 in (90 cm)
Lifespan:	20 - 24 years
Reproduction:	ovoviviparous, 4 - 40 pups per litter
Color:	blackish brown

Features: small stout body with small fins with white spines, large green oval eyes, second dorsal fin larger than the first

Habitat: bottom or edge of the continental shelf

Distribution: western north Atlantic and south Atlantic oceans

Did you know?
- You can only find the Black dogfish in the Atlantic Ocean.
- It is a small shark that lives in deep waters, so it's rare to see or encounter one.
- It may be luminescent and emit light.
- They are not abundant enough to support commercial fisheries.
- They are not dangerous to humans.
- Their predators include large fish and other large sharks that live in the same habitats and depths.

BLACKNOSE SHARK

Status: near-threatened
Maximum size: 6.5 ft (2 m)
Lifespan: 10 years
Reproduction: viviparous, 4 pups per litter
Color on top: pale green or yellow grey
Color underneath: white or yellowish

Features: Dark tip on the end of the snout, large eyes, first dorsal fin starts behind pectoral fins
Habitat: continental shelves over sandy bottoms and coral
Distribution: western Atlantic Ocean from the southern USA to the south of Brazil

Did you know?
- The Blacknose shark is cooked in a Mexican fish dish called 'pan de cazon.'
- It is often displayed in public aquariums.
- In nature, it swims fast and in groups with other Blacknose sharks, often preying on schools of small fish.
- Blacknose sharks from the Gulf of Mexico breed every year, and those from the Northwestern Atlantic breeds once every two years.

BLACKTIP REEF SHARK

Status: near-threatened
Maximum size: 5.2 ft (1.6 m)
Lifespan: 20 years
Reproduction: viviparous, 10 pups per litter
Color on top: light brown or bronze
Color underneath: white

Features: short blunt rounded snout, black and white tips on the first dorsal and lower fins, white band on flank
Habitat: shallow coastal waters, estuaries and river mouths
Distribution: western Pacific and Indian Oceans

Did you know?
- The Blacktip reef shark is the most common shark in the Indo-Pacific region.
- At birth, pups measure about 13 - 20 in (33 - 52 cm).
- It is a very social shark and likes to swim in large groups with other Blacktip reef sharks and hunt together.
- The female shark can reproduce if male sharks are not around!
- It is acrobatic and often leaps above the water, turning 3-4 times before falling back into the water.
- It is hunted for shark fin soup, shark oil, and liver.
- It scares easily, so when it is scared, it will form an S-shape and roll from side to side.

BLIND SHARK

Status: least concern
Maximum size: 3.9 ft (1.2 m)
Lifespan: 25 years
Reproduction: viviparous, 8 pups per year
Color on top: brown or black, light spots and 11 dark saddles on the back
Color underneath: yellow

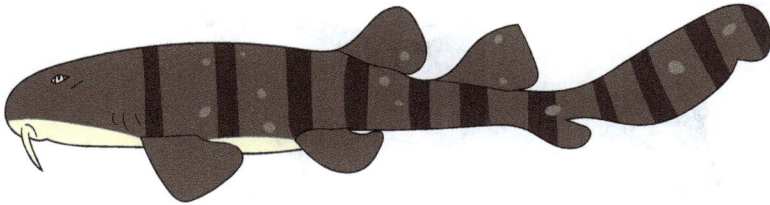

Features: wide slightly flattened head, small eyes and a nasal barbel protruding from each nostril; 2 dorsal fins close to each other at the back of the body
Habitat: shallow coastal waters
Distribution: south Queensland to southern New South Wales, Australia

Did you know?
- It has perfect vision, but it is named the blind shark as it closes its eyes when caught by fishermen.
- It hides in caves during the day and feeds at night.
- It is harmless to humans.
- Commercial fishers do not target it as it has a strong ammonia-like taste.
- It can live for 18 hours outside of water.

BLUE SHARK

Status: near-threatened
Maximum size: 12.5 ft (380 cm)
Lifespan: 20 years
Reproduction: viviparous, 15 - 30 pups per litter
Color on top: blue
Color underneath: white

Features: long conical snout, large eyes, long narrow pectoral fins in front of the first dorsal fin
Habitat: open ocean, off continental shelves
Distribution: worldwide, temperate and tropical oceanic waters

Did you know?
- The Blue shark is a slim, graceful shark.
- It swims on the surface with their fins out of the water.
- It occasionally eats seabirds from the surface of the ocean.
- It migrates depending on the availability of prey.
- The Blue shark frequently dives into the deep water to cool down, and then it returns to the surface.
- It is not a very aggressive shark, but you should approach it with caution.

BLUEGREY CARPETSHARK

Status: vulnerable
Maximum size: 33.5 in (85 cm)
Lifespan: unknown
Reproduction: ovoviviparous
Color on top: black and white when young changing to brown as an adult
Color underneath: white

Features: wide, flattish head with eyes on top, a blunt snout and a pair of barbels, large pectoral fins and dorsal fins at the very back of the body
Habitat: shallow inshore waters
Distribution: northeastern Australia

Did you know?
- The Bluegrey carpet shark is also known as Colclough's shark, blue-grey catshark, bluegrey shark, or southern blind shark.
- It closes its eyes when taken out of the water.
- There were only 9,999 adult sharks as at May 2015. The number was thought to be decreasing.
- This shark is harmless to humans and is occasionally captured by commercial fisheries on accident.

BLUNTNOSE SIXGILL SHARK

Status:	near-threatened
Maximum size:	16 ft (4.8 m)
Lifespan:	80 years
Reproduction:	ovoviviparous
Color on top:	grey, olive green or brown with a light-colored stripe along each flank
Color underneath:	lighter grey

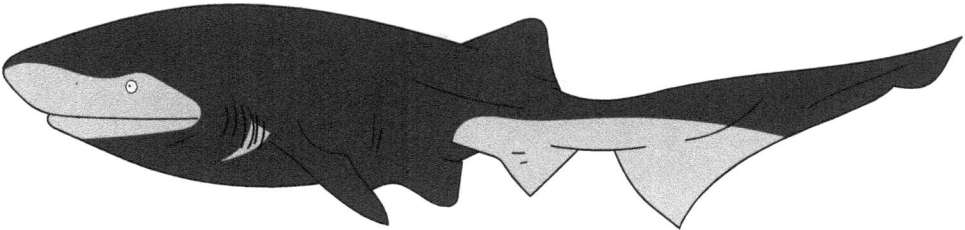

Features: Six pairs of long gill slits on either side of their broad head, single dorsal fin and a long tail, bright green eyes
Habitat: deep waters
Distribution: worldwide tropical and temperate waters

Did you know?
- The Bluntnose Sixgill shark is one of the largest sharks in the world.
- It looks like something out of the dinosaur era.
- It has six pairs of gills. Most other sharks have five pairs of gills.
- It is nocturnal, which means it is active and hunts at night. During the day, it rests on the bottom of the ocean coming up to the surface only at night so you may not see this shark unless you're swimming around in the ocean at night!

BONNETHEAD

Status: least concern
Maximum size: 59 in (150 cm)
Lifespan: 18 years
Reproduction: viviparous, 4 - 14 pups per litter
Color on top: grey to greyish brown with an occasional green tint
Color underneath: white

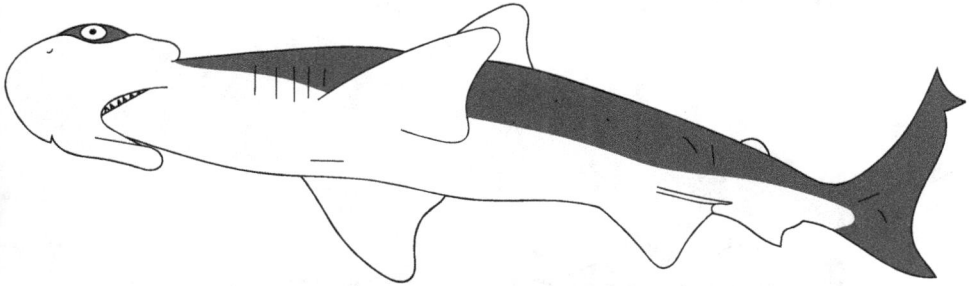

Features: shovel-shaped head
Habitat: shallow coastal waters
Distribution: warm waters of the Northern Hemisphere close to the US, Mexico, and Brazil

Did you know?

- It is the smallest type of hammerhead shark compared to the Great hammerhead that can grow to about 20 ft (6 m) long.
- It is also called a shovelhead shark or a bonnet hammerhead shark.
- It prefers the warmer waters so swims towards the equator in winter and then back to the higher latitudes in summer.
- They hang together in small schools of 10 - 15 sharks, but when they migrate, there could be hundreds or thousands of them swimming together.
- It can see more than other sharks as its eyes are at the ends of their flattened head.
- The females can store sperm for up to 4 months before fertilizing the eggs. This is so that their pups are born at the time that gives them the best chance of survival. It also has the shortest pregnancy period (4 months) of all the sharks.

BRAMBLE SHARK

Status:	near-threatened
Maximum size:	10.1 ft (3.1 m)
Lifespan:	25 years
Reproduction:	ovoviviparous, 15 - 24 pups per litter
Color on top:	dark olive or purple to dark grey or black with metallic reflections
Color underneath:	pale brown, grey or white

Features: a thick cylindrical body without the classic dorsal fin, small fins with asymmetrical tail fin and no anal fin

Habitat: deep waters

Distribution: tropical and temperate waters worldwide, mainly eastern Atlantic and western Indian Oceans

Did you know?
- In South Africa, the Bramble shark's liver oil is an ingredient in medicine.
- It is a rare deepwater slow-moving, bottom-dwelling shark.
- It only swims to the surface when it's cold.
- It is a loner and likes to swim alone.

BROADNOSE SEVENGILL SHARK

Status: near-threatened
Maximum size: 10 ft (3 m)
Lifespan: 50 years
Reproduction: ovovivaparous, 60 - 108 pups per litter
Color on top: brownish grey with small black and white spots
                         covering the back and fins
Color underneath: white or light grey

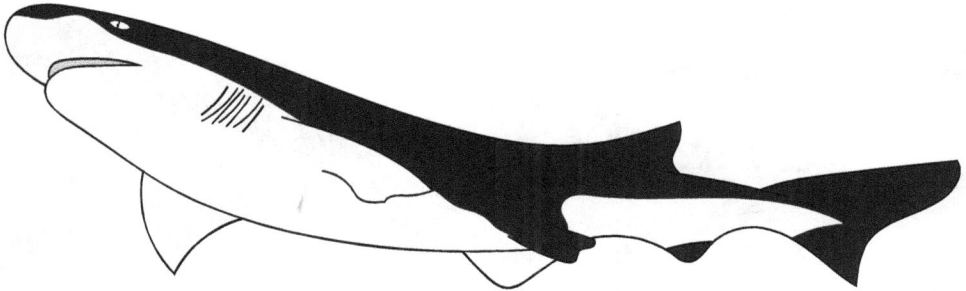

Features: Seven pairs of gills, a short blunt snout, one dorsal fin towards the back of the body
Habitat: coastal waters less than 165 ft (50 m) deep, prefers rocky, sandy or muddy bottoms and sand
Distribution: worldwide temperate waters

Did you know?
- It is in the Guiness Book of records for having the most gill sets (seven) for a shark. It shares this record with the Sharpnose sevengill shark.
- It likes to swim slowly along the bottom of the sea and then ambushes its prey with a quick burst of speed.
- It doesn't just eat fish. It also preys on dolphins, seals, other sharks, and rays.
- Fishers hunt them for their liver oil, leather, and food.

BRONZE WHALER SHARK

Status: near-threatened
Maximum size: 9.6 ft (295 cm)
Lifespan: 25 - 30 years
Reproduction: viviparous, 15 pups per litter
Color on top: bronze grey or olive green
Color underneath: white

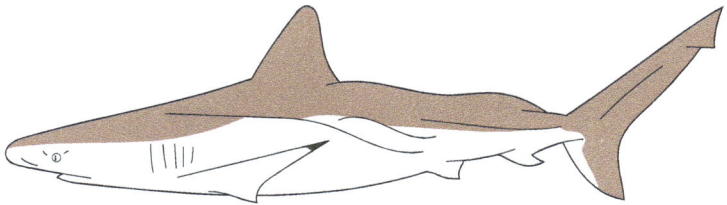

Features: a pointy snout and large pectoral fins with pointed tips
Habitat: prefers waters 330 ft (100 m) deep
Distribution: worldwide temperate waters of Indo Pacific and Atlantic Oceans and the Mediterranean Sea

Did you know?
- Fishers target the Bronze whaler shark for its meat.
- It is known to be potentially dangerous to humans, so avoid it if you can.
- It's a large shark that likes to hunt in large groups with other Bronze whaler sharks.
- This shark is sometimes confused with other similar looking sharks such as the dusky shark, blacktip shark, sandbar shark, and spinner shark.

BROWN SHYSHARK

Status: vulnerable
Maximum size: 2.3 ft (69 cm)
Lifespan: 22 years
Reproduction: oviparous
Color on top: brown with small white or black spots
Color underneath: white

Features: broadhead with very large nostrils
Habitat: inshore in shallow sandy waters
Distribution: the western Indian Ocean and the South Atlantic Ocean, off the coast of South Africa in the western Atlantic Ocean

Did you know?
- When the brown shyshark is scared, or it has been caught, it curls up and uses its tail to cover its eyes!
- Its favorite food is lobster and bony fish.
- The brown shyshark is also known as plain happy!
- This shark mainly feeds on bony fishes and sometimes lobsters.

BROWNBANDED BAMBOO SHARK

Status:	near-threatened
Maximum size:	41 in (104 cm)
Lifespan:	25 years
Reproduction:	oviparous
Color:	light brown (the young also have dark bands and spots across the body)

Features: long slender shark with lobed fins and a thick long tail, the mouth is closer to the eyes than its snout
Habitat: coral reefs and sandy and muddy bottoms
Distribution: Asia, northern Australia, and Western Australia

Did you know?
- You might want one of these in your home aquarium as it survives very well. If not, you might see one of these in a public aquarium in Australia, Europe, Mexico, Canada, or the US as they are quite popular.
- It's good eating, so fishers from India, Thailand and other southeast Asian countries target these sharks for their meat.
- They lay long eggs! They can reproduce in captivity, and their eggs are about 5.9 in (15 cm) long.

BULL SHARK

Status: near-threatened
Maximum size: 10.6 ft (3.24 m)
Lifespan: 13 - 17 years
Reproduction: viviparous, 13 pups per litter
Color on top: grey with a faint white bank on its flank
Color underneath: pale cream

Features: stocky, heavy shark with a short flat snout and small eyes, the first dorsal fin is large and triangular
Habitat: shallow coastal waters at a depth of 100 ft (30 m), seawater or freshwater
Distribution: worldwide tropical and subtropical waters

Did you know?
- The Bull Shark is known as the "pitbull of the sea." It is a large, aggressive shark, one of the top three sharks that will attack humans (the other two are the Great White Shark and the Tiger Shark).
- When it bites, it makes more than one bite, which makes it hard to repair the wound.
- It can live in seawater or freshwater. Only having 50% salt concentration in their blood helps them to survive in freshwater.
- It produces 20% more urine when it is swimming in freshwater.
- It was the inspiration for the JAWS book and movie!Its favorite food is lobster and bony fish.

BURMESE BAMBOO SHARK

Status:	data-deficient
Maximum size:	22.6 in (57.5 cm)
Lifespan:	25 years
Reproduction:	oviparous
Color on top:	brown
Color underneath:	white

Features: a long distinctive snout and thin fins, the tail is longer than the rest of the body, large openings (spiracles) behind the eyes
Habitat: tropical shallow waters
Distribution: northeastern Indian Ocean off Myanmar (Burma)

Did you know?
- The Burmese bamboo shark is very rare. Only one had ever been caught off the coast of Burma in 1963. The specimen is displayed in the Smithsonian Institute of the National Museum of Natural History in Washington DC.
- It is a small shark that swims on the bottom of the ocean.
- Not much is known about this shark, but it is related to the Carpet and Bamboo shark families.

CARIBBEAN REEF SHARK

Status: near-threatened
Maximum size: 9 ft (2.95 m)
Lifespan: 22 years
Reproduction: viviparous, 4 - 6 pups per litter
Color on top: grey-brown to olive
Color underneath: white to yellow

Features: classic shark shape with a short blunt round snout, large narrow pectoral fins, and a small first dorsal fin
Habitat: bottom dweller of continental shelves often found on coral reefs next to drop-offs
Distribution: Western Atlantic from Florida to southern Brazil, Bahamas, Gulf of Mexico and the Caribbean Sea

Did you know?

- The Caribbean reef shark is the most common shark on the coral reefs in the Caribbean Sea. It is the top of their food chain or the apex predator where they swim.
- It 'sleeps' in caves and lies motionless on the bottom of the ocean.
- In the US, you are not allowed to catch this shark commercially.
- It doesn't tend to attack people unless it feels it's being threatened. If it starts to swim upward or spiral around, quickly turns and moves, and you see its back arching like a cat's, you know it's about to attack!

CARIBBEAN ROUGHSHARK

Status: data-deficient
Maximum size: 1.7 ft (50 cm)
Lifespan: unknown
Reproduction: ovoviviparous
Color: light grey with dark bands and blotches spread over its head, body, and tail with lighter colors over pectoral and pelvic fins

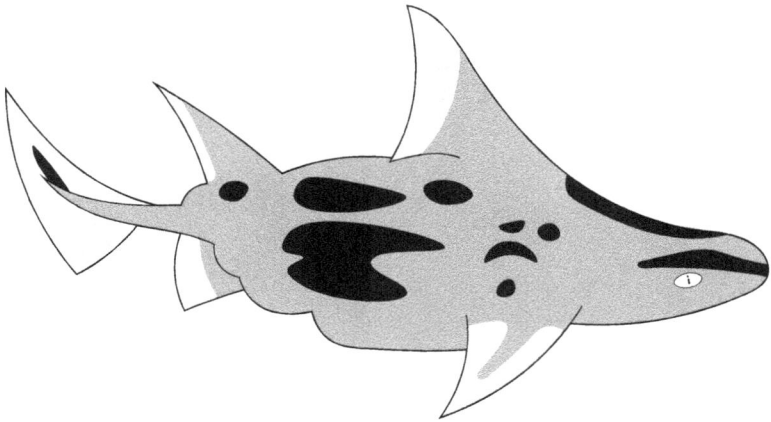

Features: short blunt snout, short body with two high dorsal fins with sharp spines
Habitat: upper continental slopes
Distribution: western Atlantic Ocean, Gulf of Mexico and the Caribbean Sea

Did you know?
- It's a chubby looking shark and quite small.
- We don't know much about this rare shark that prefers to swim in deep water. They sometimes get caught up accidentally in fishing nets.
- This shark has little to no commercial uses.
- It is suspected that the Caribbean roughshark feeds on invertebrates and fishes around their habitats.

CHAIN DOGFISH

Status: least concern
Maximum size: 18.9 in (48 cm)
Lifespan: 20 years
Reproduction: oviparous
Color on top: reddish-brown with dark brown or black chain pattern
Color underneath: light yellow

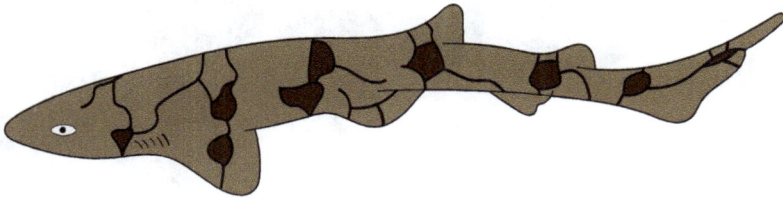

Features: chain pattern all over the body, reddish-green eyes
Habitat: rocky bottoms of continental shelves and upper slopes
Distribution: the western Atlantic Ocean from Massachusetts to Florida and the northern Gulf of Mexico to Nicaragua

Did you know?
- The Chain dogfish is also called the Chain catshark.
- This is a pretty fish and a small one, so it is a very popular fish for aquariums. But make sure you have enough room in your aquarium for the size of an adult shark.
- The male shark can bite a female four times in 30 minutes when courting her.
- A female shark has been known to store sperm for up to 843 days (28 months)!
- The baby Chain dogfish wriggles all over the place are very uncoordinated at first, bumping into everything and anything.

COOKIECUTTER SHARK

Status: least concern
Maximum size: 22 in (56 cm)
Lifespan: unknown
Reproduction: ovoviviparous, 6 - 12 pups per litter
Color on top: dark brown with a dark collar around its gills
Color underneath: lighter brown

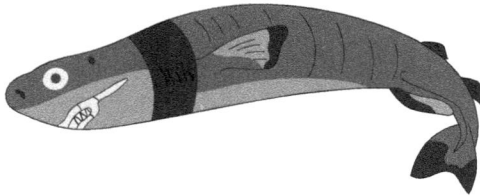

Features: small cigar-shaped body with a short conical snout and unique suctioning lips and large green eyes
Habitat: deep water
Distribution: temperate and tropical waters of the Atlantic and Pacific Oceans

Did you know?
- Not only does it lose all its teeth at once, but it also swallows the entire bottom row of 25 - 31 teeth for calcium!
- It is found mainly in deep water and swims up vertically at night to hunt.
- It glows! The bottom of the shark has photophores, which gives it a greenish glow. It is often called the luminous shark.
- Their glow attracts prey, and just before the prey can take a bite, the cookiecutter shark attaches itself to the prey with its sucking lips and sharp upper teeth. It then spins around and removes a cookie-shaped bite from the prey using its lower teeth.
- It keeps glowing for about 3 hours even after it's dead.

COPPER SHARK

Status: near-threatened
Maximum size: 11.6 ft (3.5 m)
Lifespan: 30 years
Reproduction: viviparous, 24 pups per litter
Color on top: olive green to bronze
Color underneath: white

Features: long torpedo-shaped shark with dark markings along the edges of the fins, upper teeth have outward triangular, hooked shape
Habitat: coastal areas, salt or fresh water or shallow bays
Distribution: temperate waters worldwide

Did you know?
- The copper shark is also called the bronze whaler and cocktail shark.
- It swims fast and likes to hunt in large groups.
- Commercial fishers target it for their meat, fins, oils, and skin. People also try to catch these fish as a gamefish.
- Often found within the surf zone of a beach and can leap out of the water.

CRESTED BULLHEAD SHARK

Status:	least concern
Maximum size:	5 ft (1.5 m)
Lifespan:	12 years
Reproduction:	oviparous
Color on top:	light tawny brown with large dark blotches or bands over its head, body, and tail
Color underneath:	same as the top color

Features: flattish broad-headed shark, piggish snout and high ridges over the eyes, high dorsal fins

Habitat: rocky reefs and sea bottoms

Distribution: the east coast of Australia

Did you know?

* This shark is a bottom dweller and a slow swimmer.
* It is also called the Crested horn shark or Crested Port Jackson shark.
* The Bullhead and Horn Sharks are its families.
* Its favorite food is Port Jackson shark eggs!
* It has almost no recreational or commercial use and is not dangerous towards humans.

DUSKY SHARK

Status: vulnerable
Maximum size: 11.9 ft (3.65 m)
Lifespan: 40 - 50 years
Reproduction: viviparous
Color on top: grey to bluish-grey
Color underneath: white

Features: short broadly rounded snout, no markings on fins
Habitat: continental shelves, coastal waters from the surf zone to deeper waters up to 1,312 ft (400 m)
Distribution: warm temperate and tropical waters

Did you know?
- The Dusky shark is one of the oldest sharks in the world.
- It makes good shark fin soup, so fisherman catches them for their fins as well as their meat. More than 750,000 dusky sharks are caught for the shark fin trade every year.
- It can potentially be dangerous to people as it is a large shark that swims in shallow coastal waters.
- Female sharks only give birth once every three years, so it is slow to reproduce.

EPAULETTE SHARK

Status: least concern
Maximum size: 42.1 in (107 cm)
Lifespan: 20 - 25 years
Reproduction: oviparous, 20 eggs per year
Color: creamy or brown with many small dark spots over its body (not the fins) and two large black spots with white margins above the pectoral fins

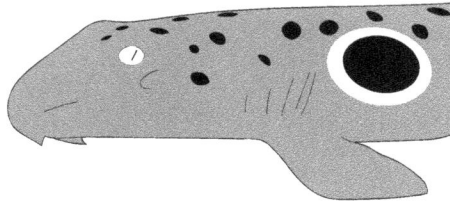

Features: small, slender shark with a short snout, a groove connecting its mouth to its nostrils and small barbels, round paddle-like pectoral and pelvic fins

Habitat: shallow water coral reefs

Distribution: the western Pacific Ocean mostly around New Guinea and northern Australia; also off Malaysia, Indonesia and Solomon Islands

Did you know?

- It can walk by moving its fins to a 90-degree angle to its body and uses them like its 'feet'. The Epaulette shark likes to look for food in tidal pools, and when the tide goes out, it gets stuck in the shallow water. When it gets stuck in tidal pools, it simply 'walks' out!
- It is closely related to the speckled carpet shark, but it doesn't have the curved dark spots that the speckled shark has.
- It is called the Epaulette shark as its two black spots look like an epaulet on a military uniform.

FRILLED SHARK

Status: least concern
Maximum size: 6.6 ft (2 m)
Lifespan: 25 years
Reproduction: ovoviviparous, 6 pups per litter
Color: brown

Features: frilly appearance of 6 pairs of gill slits, a big mouth that continues to the back of the head, 25 rows of 300 triangular shaped needle-sharp teeth
Habitat: deep waters
Distribution: Atlantic and Pacific Oceans around Australia, New Zealand, SE Asia, West Africa, Chile or the Caribbean

Did you know?
- The Frilled shark is nicknamed the living fossil as it hasn't changed much since prehistoric times.
- It is snakelike and shaped like an eel but with fins like a shark.
- It swims with its mouth open, like the Basking shark.
- A female can be pregnant for up to 3.5 years! It's a slow, growing shark.
- It has been accused of being the Loch Ness Monster or the sea serpent.
- It has been caught in really deep water at 5,150 ft (1,570 m).
- To catch its prey, it curls up like a snake and then lunges forward. It can swallow its prey in one piece!

GALAPAGOS SHARK

Status:	least concern
Maximum size:	12.1 ft (3.7 m)
Lifespan:	24 years
Reproduction:	viviparous
Color on top:	brown grey
Color underneath:	white

Features: a slender shark with a long rounded snout, large circular eyes, large pectoral fins, dual brownish dorsal fins white at the base

Habitat: inshore and offshore near or on continental shelves up to 590 ft (180 m) deep

Distribution: circumtropical around oceanic islands such as the Galapagos Islands

Did you know?
- The Galapagos shark is a large, aggressive shark.
- It resembles the grey reef shark but has a more slender body and a rounded tip on the first dorsal fin.
- It is one of the fastest hunters of the sea and usually hunts in groups.
- It's a curious nosy shark. If it finds something interesting, like a diver or a boat, it will bump its nose into them.

GHOST SHARK

Status: least concern
Maximum size: 49 in (125 cm)
Lifespan: 15 years
Reproduction: oviparous
Color: silvery-white with occasional dark markings behind the eyes or on the fins

Features: very large high set eyes and club-like structure at the end of the snout, long spine in front of the first dorsal fin
Habitat: continental shelves of cool temperate areas up to 656 ft (200 m) deep
Distribution: southern Australia and New Zealand

Did you know?
- It's often caught and sold as a silver trumpeter or white fish fillets and 'fish and chips' in Australia and New Zealand.
- It's very popular and highly valued by the Ngai Tahu Maori tribe of New Zealand.
- It's also called a chimera or a ratfish.
- The males have their sex organs on the top of their heads!
- Females can get pregnant and store sperm for later.

41

GOBLIN SHARK

Status: least concern
Maximum size: 20 ft (617 cm)
Lifespan: 35 years
Reproduction: ovoviviparous
Color: pinkish white or grey with bluish-grey fins

Features: long head with long overhanging flattened snout, small eyes and a large mouth with fanglike teeth, flabby body with short rounded fins
Habitat: outer continental shelves and seamounts
Distribution: Atlantic, Pacific, and Indian Oceans, commonly near Japan

Did you know?
- The pink color of this shark is not because their skin is pink but because it's transparent, and we can see its blood vessels.
- This is a rare shark, so we don't know much about it. We do know it's a slow swimmer, and it has poor eyesight. So to catch its food, it drifts along quietly without making too many movements, and when it gets close enough to its prey, it just sticks its jaw out to catch it.
- It has been caught deep in the ocean at 4,300 ft (1,310 m).
- It feeds off the bottom of the ocean and unfortunately has been found with human-made objects in its stomach.

GREAT HAMMERHEAD SHARK

Status: endangered
Maximum size: 20 ft (6 m)
Lifespan: 25 - 35 years
Reproduction: viviparous, 6 - 42 pups per litter
Color on top: dusky brown to light grey
Color underneath: cream

Features: flat hammer-shaped head with a prominent notch at the midline, very tall and recurved first dorsal fin
Habitat: warm coastal shallow and deep waters
Distribution: circumtropical

Did you know?

- This is the largest of the nine species of hammerhead sharks.
- Their eyes are located on the sides of its 'hammer.' For this reason, they have a big blind spot directly in front of their nose.
- Their fins are very valuable, and it's sad that fishers catch them for their fins and then throw the rest of the shark back into the water to die (as it can't swim without its fins).
- The longest known hammerhead was 20 ft (610 cm), about the height of an adult giraffe.
- Their favorite food is stingray.
- They are cannibalistic and will eat their species when they can't find other prey.

GREAT WHITE SHARK

Status: vulnerable
Maximum size: 23 ft (7 m)
Lifespan: 70 years
Reproduction: viviparous
Color on top: usually slate grey but can also be dark blue, brown or black
Color underneath: white

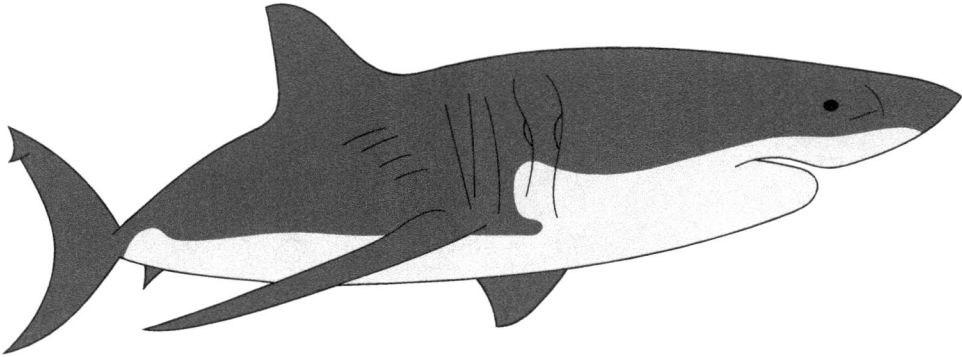

Features: classic shark shape with several rows of up to 300 jagged teeth
Habitat: coastal and outskirts of shore waters
Distribution: worldwide, mostly near Australia, South Africa, California, and the northeast US

Did you know?
- The Great White shark is the most dangerous shark in the ocean.
- About 1/3 - 1/2 of all shark attacks every year are from Great White sharks, but they are not necessarily fatal.
- It is also known as the White Shark, White Pointer, and White Death.
- It's a very clever hunter and will often surprise their prey from below. It also rolls its eyes back when it is being attacked to protect them.
- The larger pups inside the mother shark's womb will kill and eat the underdeveloped ones before they're born!
- If they're competing for the same prey with other Great White sharks, they have a tail slapping contest. The Great White shark that slaps the most gets the prey!

GREENLAND SHARK

Status: near-threatened
Maximum size: 21 ft (6.5 m)
Lifespan: 400 years
Reproduction: ovoviviparous
Color: black, brown or grey, sometimes with dark lines or spots along its back and sides

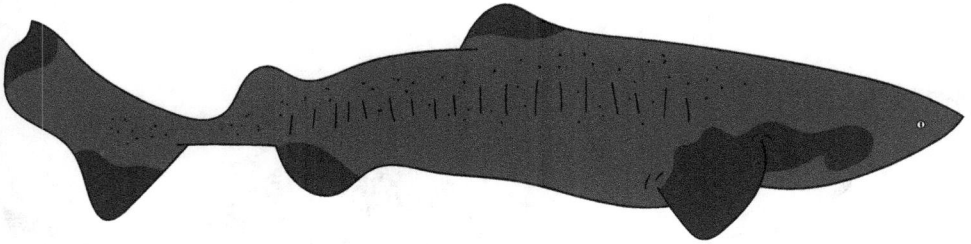

Features: large, heavyset shark with a short rounded snout, small eyes, and small dorsal fins
Habitat: up to depths of 3,937 ft (1,200 m) in freezing waters between 28 - 44 deg F (-2 - 7 deg C)
Distribution: northern Atlantic and Arctic regions

Did you know?
- It is one of the largest sharks in the world. It is also the longest-lived vertebrate on Earth.
- It grows slowly as it lives in freezing waters.
- The Greenland shark is the apex predator of icy waters, but it is a slow swimmer, so it tends to ambush its prey or scavenge for food.
- Most Greenland sharks are blind as they have parasites hanging from their eyes.
- They hunt in the dark under the ice, but this doesn't matter as they're blind!
- Their top teeth, which are thin, pointy, and small, are completely different from their bottom teeth, which are larger, broader, and smoother.

GREY REEF SHARK

Status: near-threatened
Maximum size: 8.3 ft (2.55 m)
Lifespan: 25 years
Reproduction: viviparous, 6 pups per litter
Color on top: dark to bronze grey with a black band on the back of the caudal fin
Color underneath: white

Features: large classically shaped shark with a long broad rounded snout and black-edged caudal fin
Habitat: coral reefs and near drop-offs to the open sea, also shallow lagoons next to strong currents
Distribution: the Indian Ocean and Western Pacific Ocean towards the Hawaiian Islands

Did you know?
- The Grey reef shark is considered the most dominant shark in the coral reefs.
- It is an inquisitive social shark that swims in groups during the day and hunts alone at night.
- If you see this shark start to turn its head side to side, arch its back, lowers its pectoral fins, and swim in a figure 8 loop, watch ou! This is called a threat display, and the shark may attack if it feels threatened.

HORN SHARK

Status: least concern
Maximum size: 3.3 ft (1.2 m)
Lifespan: 12 - 25 years
Reproduction: oviparous, two eggs every two weeks from February to April
Color on top: dark to light grey or brown with dark brown or black spots
Color underneath: yellow

Features: short blunt head with high ridges over the eyes, spines on dorsal and anal fins

Habitat: continental shelves on rocky reefs preferring 70 deg F (20 deg C) waters

Distribution: warm temperate and subtropical regions of eastern Pacific Ocean, off western North America

Did you know?
- The Horn shark is related to the bullhead shark.
- They move slowly on the bottom of oceans and hunt at night.
- They're homebodies, so after scavenging for food, they return to their home every time.
- You can see them in many aquariums. They have sharp spines on their fins, so be careful when you touch one.
- Their bite is the strongest compared to any other shark (relative to its size). They can eat hard-shelled creatures like crabs.
- Their teeth are purple as they love eating purple sea urchins.

JAPANESE SAWSHARK

Status: data-deficient
Maximum size: 4 ft 6 in (1.36 m)
Lifespan: 15 years
Reproduction: ovoviviparous, 12 pups per litter
Color: brown

Features: long slender body with long flat snout (saw) with 24 - 45 teeth on either side and a pair of whisker Barbels halfway on the snout
Habitat: sandy or muddy sea bottoms up to 2,670 ft (800 m)
Distribution: Northwest Pacific region of Asia - Japan, northern China, Taiwan and Korea

Did you know?
- This fish is a swimming saw! Its long nose makes up 20% of its entire body.
- The barbels make it look like it has a long thin mustache.
- It looks like a sawfish, but you can tell it's not as a sawfish doesn't have the mustache (barbels).
- Their fins are covered with scales to protect them from predators.

JAPANESE WOBBEGONG

Status: data-deficient
Maximum size: 3.3 ft (1 m)
Lifespan: unknown
Reproduction: ovoviviparous, 20 pups per litter
Color: light sandy brown with patterns of lines and darker 'saddle' markings

Features: broad flat head, skin growths around its mouth with barbels, gill slits larger than its eyes
Habitat: temperate and tropical waters of western Pacific Ocean
Habitat: sandy, rocky surfaces and coral reefs

Did you know?
- It's a type of carpet shark. Twelve different types are belonging to the Wobbegong family.
- Wobbegong' means 'shaggy beard' in the Australian Aboriginal language. It rhymes with 'hop-a-long.'
- It likes to play, hide, and seek! Due to its color, it camouflages easily on the seafloor.
- It's not a very fast swimmer, so it will sneak along the bottom and then ambush their prey when it gets close enough.
- It can walk! Using their fins, they can walk out of a shallow pool of water into another one.
- Aquariums like to show them off.

LEAFSCALE GULPER SHARK

Status: vulnerable
Maximum size: 5 ft (1.5 m)
Lifespan: 70 years
Reproduction: ovoviviparous, 5 - 8 pups per litter
Color: brown or grey

Features: large green eyes, long snout, tiny denticles or v-shaped scales on its skin, long and low first dorsal fin, no anal fin

Habitat: depths of 459 - 4,100 ft (140 - 1,250 m) on continental slopes or in open water

Distribution: northwestern and north and southeastern Atlantic, western and eastern Indian Ocean, south and northwestern Pacific and southeast Pacific

Did you know?
- The Leafscale Gulper shark is closely related to the dogfish family.
- It may be the longest living shark species that's ever existed.
- They're great hunters and come out at night.
- Fishers catch them for their meat and liver oil.

LEMON SHARK

Status: near-threatened
Maximum size: 11 ft (3.5 m)
Lifespan: 27 years
Reproduction: viviparous, 5 - 20 pups per litter
Color on top: pale yellow-brown to olive grey
Color underneath: pale yellowish-white

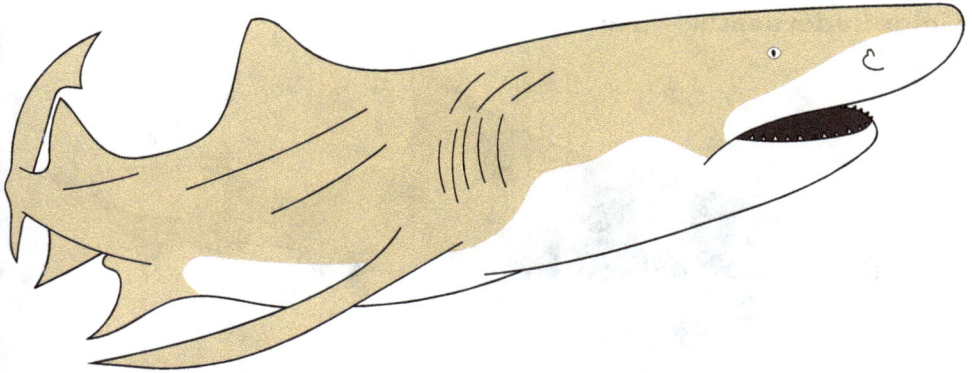

Features: 2 equal-sized triangular dorsal fins, blunt snout
Habitat: shallow waters up to 300 ft (90 m) around coral reefs, mangroves, bays, and river mouths
Distribution: western Atlantic from New Jersey to southern Brazil, eastern North Atlantic including Senegal and Ivory Coast, eastern Pacific from southern Baja to Ecuador

Did you know?

- It's yellow like a lemon. Their color helps them to camouflage well with the sandy bottoms of the ocean.
- It's a common fish seen in aquariums. This shark can handle living in aquariums for long periods, so scientists have been able to study it well.
- It's one of the fish that is caught for its fins for shark fin soup.
- It likes to eat seabirds (as well as other fish and sharks).
- It can't see very well, but it has a good sense of smell due to the magnetic sensors in their nose.
- It's a very social shark and likes swimming with their friends.
- They are gentle creatures and very much loved by shark divers.

LEOPARD SHARK

Status: least concern
Maximum size: 7 ft (2.13 m)
Lifespan: 30 years
Reproduction: ovoviviparous, 4 - 29 pups per litter
Color on top: silver or bronze grey with distinct black saddle marks and splotches on its back, sides, and fins
Color underneath: white

Features: long slim shark with a broad short snout
Habitat: sandy bottoms of bays or estuaries
Distribution: the eastern Pacific Ocean from Oregon to the Gulf of California in Mexico

Did you know?
- The Leopard shark is related to the Houndshark.
- It often gets mixed up with the Swell shark, which is reddish-brown and has a flat head.
- It breathes oxygen better than other estuary sharks as it has more red blood cells.
- They like to hang out with Smoothhounds and Piked dogfish.
- It's a very active shark and a very strong swimmer.
- It's a popular aquarium shark.
- They're homebodies and will stay in the same area for most of their lives.

MAKO SHARK

Status: endangered
Maximum size: 14 ft (4.5 m)
Lifespan: 28 - 35 years
Reproduction: ovoviviparous, 2 - 8 pups per litter
Color on top: dark slate blue or grey black
Color underneath: white

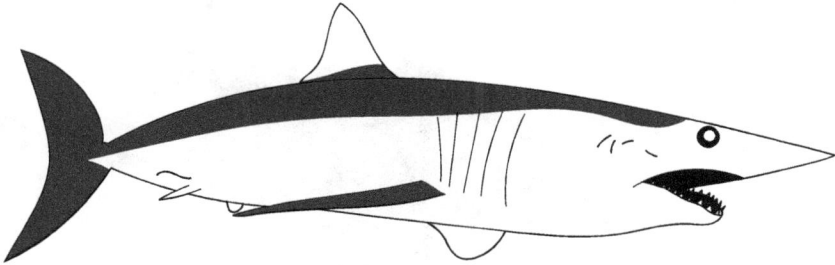

Features: visible teeth even with a closed mouth, pectoral fins as long or longer than its head, large eyes, long conical snout
Habitat: deep ocean waters
Distribution: Indian, Pacific and northwestern and eastern Atlantic Oceans

Did you know?

- It is the fastest shark in the world! The Mako shark holds the record for being the fastest shark on Earth with an average speed of 60 mph (96 kph) when hunting.
- They look mean - you can see their teeth showing even with their mouth closed. They also love jumping out of the water!
- Mako is a Maori word and the Maori people in New Zealand has been making shark tooth jewellery for years from this shark
- The Mako shark will become extinct soon if people don't stop catching them.
- In 2016, scientists tagged two mako sharks to try and predict the outcome of the US presidential election. The shark that swam the furthest would be the winner. The Donald Trump shark swam 652.44 miles (1,051 km) compared to the Hilary Clinton Shark that swam 510.07 miles (820 km). Donald Trump won!

MEGALODON SHARK

Status: extinct
Maximum size: 40 - 70 ft (12 - 21 m)
Lifespan: unknown
Reproduction: viviparous
Color: unknown

Features: unknown
Habitat: shallow waters
Distribution: worldwide warmer waters of Africa, North and South America, India, Australia, Japan, and Europe

Did you know?

- The Megalodon shark ruled the warm waters of Earth 70 - 10 million years ago.
- It became extinct about 2.6 million years ago. The information that we have on Megalodon is based mostly on teeth fossils.
- It was very much like a Great White Shark of today but meaner and much bigger.
- It was the largest carnivorous fish on Earth and may have fed on whales
- Megalodon means 'giant tooth.' Their teeth measured over 7 in (17.7 cm). The largest Megalodon tooth was found in Peru and was 7.48 in (19 cm) long.
- It had about five rows of 276 teeth in its mouth at any one time. Each tooth was three times larger than a Great White Shark's tooth.

MEGAMOUTH SHARK

Status: least concern
Maximum size: 18 ft (5.49 m)
Lifespan: unknown
Reproduction: ovoviviparous
Color on top: brownish black
Color underneath: white

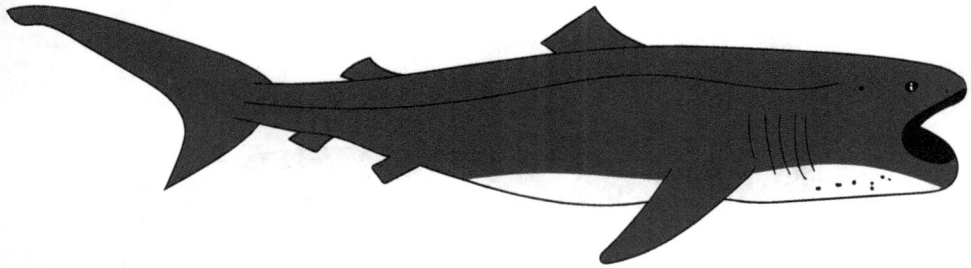

Features: large head with rubbery lips, large mouth with small teeth and silvery-white upper lip with a broad, rounded snout
Habitat: upper waters of the open ocean, up to 520 ft (160 m) deep
Distribution: Indian, Pacific, and Atlantic Oceans

Did you know?
* The Megamouth shark is so rare that scientists have kept a list of every shark that has ever been found. Up until 5 March 2018, only 99 sharks have been caught or sighted.
* It is called a Megamouth as it has an enormous mouth. It swims with its mouth open to feed on plankton and jellyfish, just like the basking shark and the whale shark.
* It is sometimes mistaken for a young orca due to its large mouth.

NECKLACE CARPETSHARK

Status:	least concern
Maximum size:	36 in (91 cm)
Lifespan:	unknown
Reproduction:	oviparous
Color on top:	dark to light brown with white spots, dark and light blotches, and dark saddles
Color underneath:	white

Features: a long tube-shaped shark with small blackish-brown fins and a short rounded snout, a wide dark collar with dense white spots (looks like a necklace)

Habitat: sandy bottoms, rocky reefs, kelp, and seagrass beds

Distribution: southern Australia

Did you know?

- It is a night fish, so you're unlikely to see it during the day.
- It also likes to hide in caves and camouflages very well on the floor of the ocean.
- Sometimes people think it is a type of catshark, but it is related more closely to a wobbegong or a nurse shark.

NURSE SHARK

Status: data-deficient
Maximum size: 15 ft (4.5 m)
Lifespan: 25 - 35 years
Reproduction: ovoviviparous, 30 pups per litter
Color on top: light tan to dark brown
Color underneath: lighter tan or white

Features: its mouth has barbels, rounded fins, very small eyes
Habitat: shallow waters
Distribution: Atlantic and eastern Pacific Oceans

Did you know?
- The Nurse shark is a homebody and will return to their preferred cave and resting place, often with other nurse shark friends, after hunting.
- It's a pest! Fishermen call them a pest when they're caught as bycatch in nets - they're often released alive. They were previously caught for their liver oil and skin but not as much now.
- It makes a strange sucking sound when it forages around for prey in the sand. It sleeps on sandy bottoms during the day and hunts at night.
- They're cannibals - their larger pups eat the underdeveloped ones.
- They're easy to catch because they swim so slowly.

PACIFIC SHARPNOSE SHARK

Status: data-deficient
Maximum size: 2.3 ft (70 cm)
Lifespan: Nine years
Reproduction: viviparous
Color on top: grey or brownish-grey to bronzy
Color underneath: white

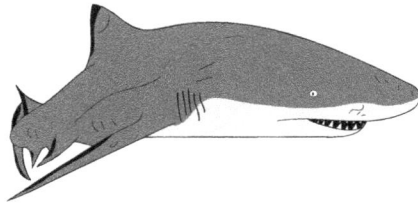

Features: a very small shark with dusky blackfin edges, a long snout, small wide-spaced nostrils, and big eyes
Habitat: shallow estuaries and coastal waters over sandy and muddy bottoms
Distribution: tropical Indo-west Pacific Ocean, often along the coast of Mexico

Did you know?
- They are abundant, but they are overfished now.
- Fishers use their meat as chunk bait to target larger sharks.
- You can often see them in an aquarium.
- Female Pacific sharpnose sharks are generally larger than the males.
- It is suspected that their diet consists mainly of crustaceans and other small fish.

PACIFIC SLEEPER SHARK

Status: data-deficient
Maximum size: 23 ft (7 m)
Lifespan: 40 years
Reproduction: ovoviviparous, 10 pups per litter
Color: grey or greyish black

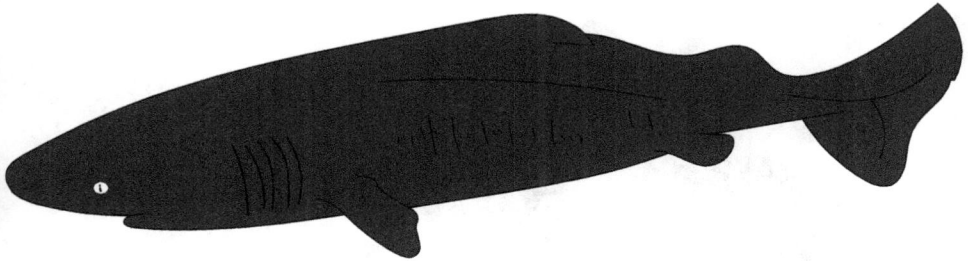

Features: wide blunt head with a short rounded snout and very small eyes, cylindrical body, low dorsal fins, and a long sweeping tail fin
Habitat: deep cold waters, up to and over 2,000 m (6,500 ft) deep
Distribution: the Pacific Ocean, including off Japan and Siberian coast, south of Tasmania, Australia, New Zealand, and Uruguay

Did you know?

- The Pacific Sleeper Shark is very big, but you probably won't see or catch one as it likes to swim in very deep waters. It also doesn't like light, so it only comes up to the surface at night.
- Its favorite meal is giant Pacific Octopus and flatfish like sole and flounder.
- It cruises around slowly and will sneak up on its prey. It sucks the prey in and then rolls its head around to swallow it. It prefers to scavenge for food than hunt.
- It has a very powerful bite, so don't put your hand in their mouth.
- Its closest relative is the Greenland shark.
- It is nearly blind because of parasites that grow on its eyes.
- In 2015 a Pacific Sleeper shark was found under an active volcano near the Solomon Islands.

PORBEAGLE SHARK

Status: vulnerable
Maximum size: 12 ft (365 cm)
Lifespan: 30 - 65 years
Reproduction: ovoviviparous
Color on top: dark grey
Color underneath: white

Features: white or light grey free rear tip of the dorsal fin, long conical snout

Habitat: mainly open ocean

Distribution: the Atlantic Ocean in the northern hemisphere from New Jersey to Canada and Greenland and the northwest coast of Africa up to Iceland, Norway, Sweden, and Russia

Did you know?

- The Porbeagle belongs to the mackerel shark family, such as the mako and white shark, but it rarely attacks people.
- It is named after its porpoise-shaped body, and it hunts like a beagle.
- It is frequently caught for its meat and oil, to make fishmeal (fertilizer) and its fins for shark fin soup. It is also a popular gamefish.
- Many fishers find them a nuisance now as they steal the fish that they've caught off their lines.
- They can swim very fast to catch their prey.
- When mating, the male shark bites the female shark to hold on to her, so she doesn't getaway!

PORT JACKSON SHARK

Status: least concern
Maximum size: 53.9 in (137 cm)
Lifespan: 30 years
Reproduction: oviparous
Color on top: light grey-brown with black bands over their eyes and across the back in the shape of a harness
Color underneath: white

Features: blunt head, small mouth, prominent crests above their eyes
Habitat: close to shore over sandy, rocky or muddy bottoms
Distribution: South coast of Australia

Did you know?
- The Port Jackson shark is the largest of its family (Heterodontid).
- It looks like an alien if you look at it front on.
- It can eat and breathe at the same time. Most sharks must swim with their mouth open to force water over their gills to get oxygen, but the Port Jackson shark can stay still and still pump water over its gills.
- When they feed, they suck in water and sand and then blow the sand out their gills so that they keep the food.
- Female Port Jackson sharks lay a pair of eggs every 10 - 14 days from August to November i.e., a total of 16 eggs every year. Unfortunately, 89% of their eggs are eaten by other sharks before they hatch.
- You can buy one for your home aquarium, and it will live for a long time. Large aquariums have been successful in breeding them.

PORTUGUESE DOGFISH SHARK

Status: near-threatened
Maximum size: 5.2 ft (1.58 m)
Lifespan: 70 years
Reproduction: ovoviviparous
Color: dark brown

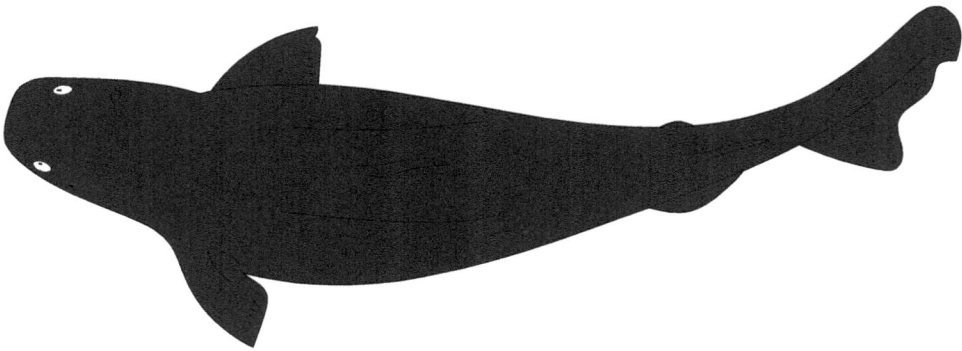

Features: shaped like a fish more than a shark with small fins set far back
Habitat: deep water
Distribution: western North Atlantic Ocean, the eastern Atlantic, western Mediterranean Sea, and the western Pacific Ocean.

Did you know?
- The Portuguese dogfish shark likes to swim way down at depths of over 3,300 ft (1,000 m). It is one of the deepest dwelling shark species.
- It has been found once at 2.3 mi (3.7 km) beneath the surface.
- It belongs to the Sleeper Shark family, which are known to be slow swimmers, but this one's the opposite - it is a fast active predator.
- Their top teeth are like spears and their bottom teeth-like blades.
- In World War II, the Japanese fished them as they found them delicious and their liver oil valuable.

PRICKLY DOGFISH SHARK

Status:	data-deficient
Maximum size:	28.3 in (72 cm)
Lifespan:	unknown
Reproduction:	ovoviviparous
Color:	brownish grey all over, dorsal fins are often translucent to white

Features: humped back body with a flat lower surface and very rough skin, two sail-like dorsal fins with spines
Habitat: 660 - 3,300 ft (200 - 1,000 m)
Distribution: Australia from central New South Wales and south to the Great Australian Bight, and New Zealand

Did you know?

- The Prickly dogfish shark has very rough skin that feels like sandpaper, so it is also called a Roughshark.
- It has a chubby appearance and a triangular-shaped cross-section.
- Its body is shaped to swim efficiently at the bottom of the ocean. A humpback shape is more energy efficient for bottom fish than fish that swim in open waters that tend to have a cigar-shaped body.
- It can be found swimming in the Twilight Zone of the ocean, which is 660 - 3,300 ft (200 - 1,000 m) just under the Sunlight zone.
- A female prickly dogfish sharks can carry 7 or 8 embryos at a time divided between a right and a left uterus!

PYJAMA SHARK

Status: near-threatened
Maximum size: 3.1 ft (95 cm)
Lifespan: unknown
Reproduction: oviparous
Color on top: grey with long stripes from head to tail
Color underneath: white

Features: small shark with long stripes along its body, short barbels, dorsal fins set back on the body
Habitat: coastal shallow waters, often in the surf or over rocky bottoms
Distribution: southeast Atlantic and western Indian Oceans often found near South Africa

Did you know?
- This shark looks like it's wearing pajamas! It's also called the Striped Catfish.
- It lazes around and hides in caves during the day and wakes up at night to feed.
- The Broadnose Sevengill shark is its biggest enemy as they swim in the same waters.
- You can see it quite often in aquariums as it's such an attractive shark.

SAILFIN ROUGHSHARK

Status: data-deficient
Maximum size: 118 cm
Lifespan: unknown
Reproduction: ovoviviparous, 7 - 23 pups per litter
Color: dark brown all over

Features: broad flat head with a short snout and large eyes, two dorsal fins with sharp spines, the first dorsal fin leans backward and is triangular, no anal fin, small circular gill slits
Habitat: deep waters
Distribution: the northeast Atlantic Ocean, off Scotland and south to West Africa

Did you know?

- The Sailfin Roughshark, also called the Kite-Fin shark, is one of the types of roughsharks. All the roughsharks belong to the Dogfish shark family.
- They differ from other roughsharks because the shape of their gill slits or spiracles is almost round. The other roughsharks have longer shaped gill slits.
- It is a rare shark that swims very slowly.
- When it dies, it changes its color to black.

SAND SHARK

Status:	vulnerable
Maximum size:	9.8 ft (3 m)
Lifespan:	30 - 35 years
Reproduction:	ovoviviparous
Color on top:	light brown
Color underneath:	greyish-white

Features: large bulky shark with flattened conical snout, broad triangular fins that are the same size and an asymmetrically shaped tail fin, humped back, long mouth extends behind the eyes with jagged teeth sticking out
Habitat: inshore, surf zone, shallow bays, coral and rocky reefs
Distribution: worldwide except the eastern Pacific Ocean

Did you know?
- You can see its teeth sticking out even when its mouth is closed.
- It is fished commercially - their meat for food, their fins for shark fin soup, and their jaws and teeth for trophies and ornaments.
- Because of their large size and fierce appearance, they are often caught and displayed in aquariums. Bertha was a female Sand Shark that lived at the New York Aquarium in Coney Island for over 40 years!
- It swallows air so that it can float on the surface of the water and then sneak up on its prey without moving too much. No other shark can do this!
- When a diver approaches too close, it thumps its tail to make a really loud boom!

SAWBACK ANGELSHARK

Status: critically endangered
Maximum size: 6.7 ft (2 m)
Lifespan: unknown
Reproduction: ovoviviparous
Color on top: blotchy grey
Color underneath: white

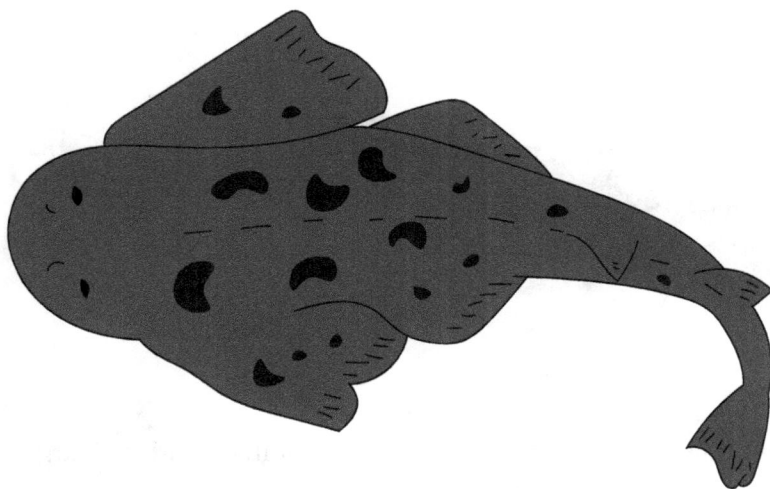

Features: large pectoral fins that spread out like wings, flat body
Habitat: offshore on muddy bottoms
Distribution: the eastern Atlantic Ocean, the western Mediterranean Sea and the coast of West Africa

Did you know?
- The Sawback Angelshark is also called a monkfish or spiny angelshark.
- They could become extinct soon. There used to be lots of Sawback Angelsharks in the Mediterranean and West Africa. However, overfishing has wiped it out in these places, and it is rare everywhere else.
- It likes to hide at the bottom of the ocean and can camouflage well.
- It can stay still for weeks until it's ready to strike at prey.

SHARPNOSE SEVENGILL SHARK

Status: near-threatened
Maximum size: 4.6 ft (1.4 m)
Lifespan: 50 years
Reproduction: ovoviviparous, 6 - 20 pups per litter
Color on top: brownish grey
Color underneath: white

Features: slender cigar-shaped body with pointed head and large green eyes, long narrow mouth, seven pairs of gills, single dorsal fin set back on its body
Habitat: deep waters
Distribution: worldwide except the eastern North Pacific Ocean

Did you know?
- The Sharpnose sevengill shark is the smallest cow shark.
- It can also be called the Perlon shark and one-finned shark.
- It has larger eyes than the only other sevengill shark, the Broadnose sevengill shark. It also has a smaller body, no black spots, and only five rows of large teeth when the Broadnose has six rows.
- Almost all sharks have five pairs of gills, and this shark has 7.
- It's a bit poisonous, so not worthwhile catching or eating.
- It is very aggressive when caught but otherwise harmless if you leave it alone.

SILVERTIP SHARK

Status: vulnerable
Maximum size: 10 ft (3 m)
Lifespan: 25 years
Reproduction: viviparous
Color on top: dark grey with bronze tint
Color underneath: white

Features: slender with a long broad rounded snout, silver-white tips, and borders on all fins
Habitat: shallow waters over coral reefs
Distribution: tropical waters in the Pacific and Indian Oceans

Did you know?
- The Silvertip Shark is a very large, aggressive shark that will fight other large sharks for food.
- It has the squarest nose of all the reef sharks.
- Its meat is considered a delicacy in South East Asia. It is heavily fished for its skin, liver oil, and meat.
- It is very dangerous to humans.
- Their best friend is the Rainbow Runner fish, who swim alongside the shark and cleans off its parasites.
- If you see a female silvertip shark without a dorsal fin, it's probably because of a male bit it off when mating.
- Try not to confuse it with the whitetip reef shark, which doesn't have the white on their pectoral fins.

SMOOTH-HOUND

Status:	near-threatened
Maximum size:	3.6 ft (1.1 m)
Lifespan:	25 years
Reproduction:	viviparous, 7 - 14 pups per litter
Color on top:	grey
Color underneath:	dirty white

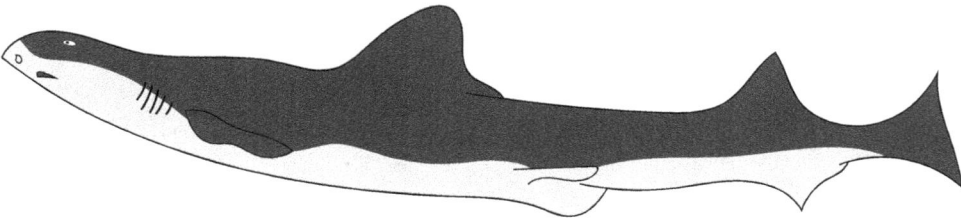

Features: large slender shark with short head and pointed snout, broad, large close-set oval eyes, long mouth, long pectoral fins, asymmetrical tail fin

Habitat: shallow sandy, muddy bottoms

Distribution: temperate east Atlantic ocean from the UK to Madeira, Angola to South Africa and Indian Ocean coast

Did you know?
- It is a shy shark that avoids people.
- It is a relative of the Dogfish shark. They come from the same family.
- In the US, you may have eaten smooth-hound as fish and chips. They're common and are caught for food and liver oil.
- They're also targeted as a popular game fish and also caught for aquariums.

SPINNER SHARK

Status: near-threatened
Maximum size: 9.8 ft (3 m)
Lifespan: 15 - 20 years
Reproduction: viviparous, 3 - 20 pups per litter every two years
Color on top: grey bronze
Color underneath: white

Features: slender shark with a black tip on all fins and pointy snout longer than the width of the mouth
Habitat: shallow waters less than 98 ft (30 m) deep
Distribution: subtropical, tropical and temperate waters of Mediterranean, Atlantic, Gulf of Mexico, parts of South America and Ino West Pacific including Red Sea, Japan, and Australia

Did you know?

- The Spinner shark spins as it swims after fish. It can then leap into the air and spin three times before it falls back into the water!
- It can leap up to 20 ft (6.1 m) into the air.
- The Spinner shark looks like a blacktip shark, but it's bigger and has a distinct black tip on their anal fin (the blacktip shark doesn't).
- It gets very excited around food and goes into a feeding frenzy.

71

TASSELLED WOBBEGONG

Status: near-threatened
Maximum size: 4 ft (1.25 m)
Lifespan: 20 - 30 years
Reproduction: ovoviviparous, 20 pups per litter
Color: light brown with a reticulated pattern of dark lines and large dark spots

Features: flat, broad head with a fringe from the tip of the snout to the base of the pectoral fins
Habitat: shallow waters near coral reefs
Distribution: the western Pacific Ocean off eastern Indonesia, Papua New Guinea, and northern Australia

Did you know?
- It is an unusual looking shark with a fringe at the front of its flat head.
- It camouflages well with the bottom of the ocean, so it is good at ambushing its prey.
- You will probably find one in US aquariums as it's really popular.

THRESHER SHARK

Status: vulnerable
Maximum size: 20 ft (6 m)
Lifespan: 22 years
Reproduction: ovoviviparous, 2 - 5 pups per litter
Color on top: dark brown or slate grey or black
Color underneath: white with dark spots near the pelvic fin

Features: very long tail or caudal fin
Habitat: prefers open deep waters but can also be found in shallower coastal waters
Distribution: Pacific and Indian Oceans, and parts of Atlantic Ocean

Did you know?
- The Thresher shark's tail is almost as long as the whole shark.
- There are three types of thresher sharks - the pelagic, bigeye, and common Thresher.
- Its closest relative is the extinct Megamouth Shark.
- They use their long tail to stun their prey and sometimes humans if they get too close!
- They can jump out of the water like a dolphin. This is called 'breaching.'
- It is also called the Fox shark. They are shy and are scared of people.
- It can swim up to 30 mph (48.2 kph). Their torpedo-shaped body and their tail make them very fast, strong swimmers.

TIGER SHARK

Status: near-threatened
Maximum size: 24 ft (7.4 m)
Lifespan: 50 years
Reproduction: ovoviviparous, 10 - 82 pups per litter
Color on top: brown, grey or black with dark grey vertical stripes or spots
Color underneath: white or pale grey or dirty yellow

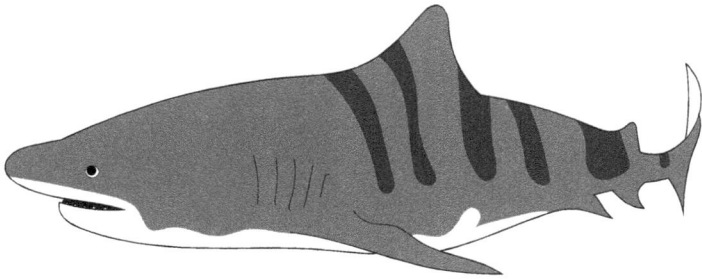

Features: dark grey stripes or spots
Habitat: prefers murky shallow coastal waters but can also be found in the open ocean
Distribution: temperate and tropical waters, except the Mediterranean Sea.

Did you know?
- It usually swims slowly and then gives a burst of speed to catch its prey. In the Great Barrier Reef, they hunt sick or dying sea turtles as they are too slow to catch the healthy ones.
- Tiger sharks, bull sharks and white sharks are often called 'The Big Three' for the number of attacks on humans. The White shark has the most records of human attacks, and the Tiger shark comes next. It has been blamed for most of the attacks in Hawaii and Australia.
- Seagrass grows better when there are tiger sharks around. This is because they scare off the turtles and dugongs who like to eat the grass.

74

WHALE SHARK

Status:	endangered
Maximum size:	60 ft (18 m)
Lifespan:	100 years
Reproduction:	ovoviviparous
Color on top:	dark grey, blue or brown with pale yellow stripes or dots
Color underneath:	white

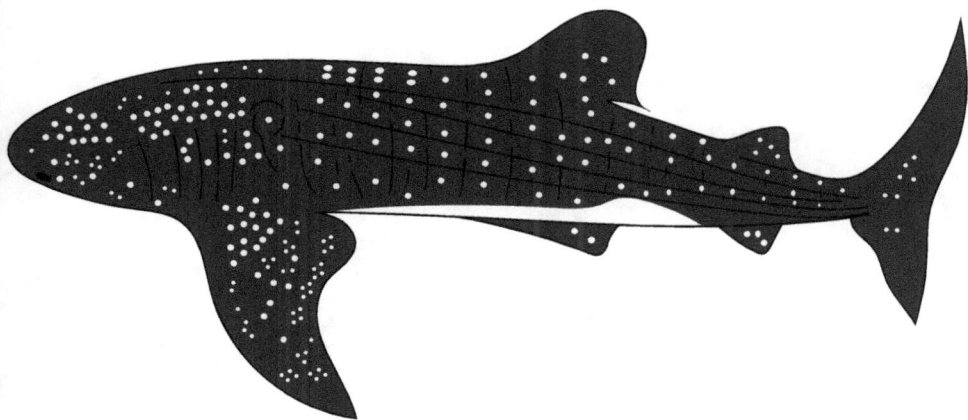

Features: back and sides have a unique checkboard pattern of spots and bars, broad head with a short snout and a mouth at the tip of the snout
Habitat: deep waters and over coral reefs
Distribution: warm waters around the equator

Did you know?

- The Whale shark is the largest fish and shark in the whole world.
- Its mouth can open up to 5 ft (1.5 m) wide. They have very small teeth.
- It's a slow swimmer with an average speed of 3 mph (5 kph).
- It is a gentle soul and completely harmless to humans so that you can swim alongside one.
- The spot patterns on each whale shark are unique, like a fingerprint.
- In springtime, they migrate to Australia to feed on plankton.
- In 1995, a female whale shark was caught pregnant with 300 embryos!

About the Author

Thank you for purchasing this book, we hope you enjoyed it!

What is your favorite shark?

The whale shark is the Professor's favorite because of its unique checkboard pattern and spots! Did you know that the Professor created this beautiful cover and all of the illustrations in this book ... and he's only eight years old! The Professor loves all things animals, cars, science, food, and dinosaurs. And with his range of books, he hopes to inspire other kids to try new things, to learn more about the world, and to be curious!

We would really appreciate it if you have a couple of minutes to leave a review on the platform that you purchased this book from - it would mean the world to us!

To claim your free printable shark poster visit bit.ly/shark-infographic or scan the QR code below.

To find our other books visit
amazon.com/author/professorsmart

To get in touch with us:
professorsmartpublishing@gmail.com

www.ingramcontent.com/pod-product-compliance
Lightning Source LLC
Chambersburg PA
CBHW050632190326
41458CB00008B/2235